Emmanuel Michael Omola
Abdullahi Kawo
Shamsuddeen Usman

Microbial & physico-chemical quality of yoghurt brands in Kano,Nigeria

Emmanuel Michael Omola
Abdullahi Kawo
Shamsuddeen Usman

Microbial & physico-chemical quality of yoghurt brands in Kano,Nigeria

Quality evaluation of Kano market yoghurt

LAP LAMBERT Academic Publishing

Impressum / Imprint
Bibliografische Information der Deutschen Nationalbibliothek: Die Deutsche Nationalbibliothek verzeichnet diese Publikation in der Deutschen Nationalbibliografie; detaillierte bibliografische Daten sind im Internet über http://dnb.d-nb.de abrufbar.
Alle in diesem Buch genannten Marken und Produktnamen unterliegen warenzeichen-, marken- oder patentrechtlichem Schutz bzw. sind Warenzeichen oder eingetragene Warenzeichen der jeweiligen Inhaber. Die Wiedergabe von Marken, Produktnamen, Gebrauchsnamen, Handelsnamen, Warenbezeichnungen u.s.w. in diesem Werk berechtigt auch ohne besondere Kennzeichnung nicht zu der Annahme, dass solche Namen im Sinne der Warenzeichen- und Markenschutzgesetzgebung als frei zu betrachten wären und daher von jedermann benutzt werden dürften.

Bibliographic information published by the Deutsche Nationalbibliothek: The Deutsche Nationalbibliothek lists this publication in the Deutsche Nationalbibliografie; detailed bibliographic data are available in the Internet at http://dnb.d-nb.de.
Any brand names and product names mentioned in this book are subject to trademark, brand or patent protection and are trademarks or registered trademarks of their respective holders. The use of brand names, product names, common names, trade names, product descriptions etc. even without a particular marking in this work is in no way to be construed to mean that such names may be regarded as unrestricted in respect of trademark and brand protection legislation and could thus be used by anyone.

Coverbild / Cover image: www.ingimage.com

Verlag / Publisher:
LAP LAMBERT Academic Publishing
ist ein Imprint der / is a trademark of
OmniScriptum GmbH & Co. KG
Heinrich-Böcking-Str. 6-8, 66121 Saarbrücken, Deutschland / Germany
Email: info@lap-publishing.com

Herstellung: siehe letzte Seite /
Printed at: see last page
ISBN: 978-3-659-40424-5

Zugl. / Approved by: Kano, Bayero University Kano, Diss.,2012

ACKNOWLEDGEMENT

I bless God for seeing me through this book. My profound gratitude goes to Prof. A. H. Kawo and Prof. M. D. Mukhtar for their intelligent criticism and useful suggestions. I am highly indebted to them.

I also wish to express my appreciation to Dr. Yahaya Mustapha. I sincerely acknowledge the technical assistance of all the staff of the Microbiology laboratory, Post graduate laboratory, Chemistry laboratory and Central laboratory of Bayero University Kano.

My gratitude goes to my wife, Helen for her care and understanding and most especially the children who often missed me while working in the laboratory.

I am also indebted to Miss Garima Nabh, and Hajia Halima Ahmed for their mutual support.

Finally, I wish to thank Francis Innocent for type-setting this work.

TABLE OF CONTENT

CONTENT	PAGE

CHAPTER THREE

LIST OF TABLES

LIST OF PLATES

LIST OF FIGURES

FIGURE   **PAGE**

CHAPTER ONE

1.0   INTRODUCTION

1.1   BACKGROUND OF THE STUDY

Yoghurt is a semi-solid fermented milk product which originated in Mesopotamia thousands of years ago. Evidence has shown that these people had domesticated goats and sheep around 5000 B.C. The milk from these animals was stored in gourds, and in the warm climate, it naturally formed a curd. This curd was an early form of yoghurt (Beel, 1994).

One legend tells that yoghurt was born by a miracle of nature. Micro-organisms of various kinds happened to land in a pitcher of milk that belonged to a Turkish nomad. The result was what the Turks called "Yogurut. The name 'yogurut' was supposedly introduced in the 8th century and was changed in the 11th century to the current version yoghurt (Bylund, 1995).

In the Bible, it is recorded that when the Patriarch, Abraham entertained three angels, he put before them soured and sweet milk (Genesis 18:8).

The oldest writings mentioning yoghurt are attributed to Pliny, the Elder, who remarked that certain nomadic tribes knew how **"to *thicken the milk into a substance with an agreeable acidity"*** (Columbia Encyclopedia, 2003).

In the early 1800s, men used yoghurt to clean their goats and sheep. Many women also used yoghurt to wash their bodies and hair. Yoghurt was the best known cleaning agent at the time (Wisegeek, 2011).

An early account of a European encounter with yoghurt occurs in French clinical history. King Francis suffered from a severe diarrhea which no French doctor could cure. His ally, Suleiman the Magnificent sent a doctor, who allegedly cured the patient with yoghurt. Being grateful, the French King spread around the information about the food which had cured him (Wisegeek, 2011). Yoghurt was first introduced to the United States in the first decade of the twentieth century, influenced by Elie Metchnikoffs *The Prolongation of Life; Optimistic Studies* (1908).

In 1919, Isaac Carasso who was from Ottoman Salonika industrialized the production of yoghurt. While yoghurt has been around for many years, it is only recently (within the last 30-40 years) that it has become popular. It is an important part of the modern day diet in many communities. Its acceptability cuts across socio-economic, religious and cultural background but it is highly vulnerable to contamination with pathogenic strains (Gadaga *et al.*, 2004).

The convenience of it as a ready - made breakfast food and the image of yoghurt as a low–fat healthy food, It is served as an appetizer, refreshing drink or as an accompaniment to a meal (Idise *et al.*, 2009).

1.2. JUSTIFICATION FOR THE STUDY

Food safety is of increasing importance in the food industry and the quality of yoghurt in the local market varies from one producer to another. The main reason is the susceptibility of the product to pathogenic strains (Dalgic and Belibagh., 2008). These have necessitated this study to ascertain their physicochemical and microbiological quality with a view to create awareness among common people about the existing situation and protect the consumers' health and rights.

1.3   AIM AND OBJECTIVES OF THE STUDY

1.3.1 AIM OF THE STUDY

This work is aimed at determining the microbial and physico-chemical quality of some yoghurt brands sold in Kano, Nigeria.

1.3.2 OBJECTIVES OF THE STUDY

The objectives of this study included the following:

1. To determine the physical characteristics (Temperature, Viscosity, Specific gravity, pH), Sensory properties and the chemical characteristics (Fat content and Titratable acidity) of yogurt samples

2. To enumerate the aerobic mesophilic bacteria, coliform bacteria and yeasts and moulds levels in the samples.

3. To isolate and characterize *Staphylococcus aureus, Escherichia coli, Salmonella* and Yeast and moulds from the samples.

2.0 LITERATURE REVIEW

2.1 CONCEPT OF FERMENTATION

Yoghurt is derived from the Turkish word "Jugurt" reserved for any fermented food with acidic taste (Younus *et al.*, 2002).

The process of fermentation was used as a means of preserving a highly perishable product and to produce new flavors for an old food staple. In the early years of milk fermentation, milk was simply allowed to be fermented by its normal microbiota, but the actual process was not completely understood. Cultures could be maintained by inoculating fresh milk with fermented milk (Kerr and McHale, 2001). Today, lactic acid-producing microorganisms are added to milk to decrease the pH of the milk and produce many different fermented milk products.

Traditionally, yoghurt is manufactured using starter cultures of selected lactic acid bacteria which are added to the milk to fulfill the desired fermentation (Mayo, 1993). These bacteria ferment the lactose in the milk to lactic acid, causing the milk to curdle and form yoghurt. If the product is not pasteurized, the result is yoghurt with "active cultures" (Sahan *et al.*, 2008). Lactic acid bacteria are fastidious microorganisms and their growth is often restricted in milk because of its paucity in essential nutrients, thus the success of milk fermentation relies most often upon the synergy between *S. thermophilus* and *L. bulgaricus*. Because both bacteria are able to grow alone in milk, this indirect positive interaction is called proto-cooperation (Courtin and Rul, 2004).

Yoghurt's popularity has grown and is now consumed in most parts of the world. Although, the consistency, flavor and aroma may vary from one region to another, the basic ingredients and manufacturing process are essentially consistent (Columbia Encyclopeadia, 2003).

In a study carried out by Taura *et al.* (2005) around the old campus of Bayero University Kano. Up to 60% of the samples had mean aerobic mesophilic bacterial counts lower than the maximum acceptable limit of 5×10^4 cfu/ml. Mold counts greater than the safe limit (1 mold/ml) were observed in 90% of the samples. Coliform counts were higher than the standard acceptable limit of 10 cfu/ml in 55% of the brands.

In a related study involving yoghurt samples obtained from retail outlets in Kaduna metropolis. No yoghurt brand had total aerobic mesophilic bacterial count up to 10.0. None of the fungal count, staphylococcal count and coliform count in all the yoghurt brands analyzed was up to 1×10^1 cfu/ml (Egwaikhide and Faremi, 2010)

In another study carried out on yoghurt stabilized with sweet potato starch in River state South-South, Nigeria. The result indicated that the coliform and fungal count of the samples were nil followed by reduction in total viable count (Okoye and Animalu, 2009).

Similarly, in a study to determine the safety of yoghurt sold in Owerri metropolis in Imo state, South-East Nigeria. The isolates were identified as *Streptococcus* and *Lactobacillus* species. No pathogen was isolated from any of the samples analysed (Oranusi *et al.*, 2011)

Also in a study to determine the bacterial population of yoghurt sold in Enugu state, Eastern Nigeria. Reports indicated that the total viable count of bacterial was in the range

5

of 1.4 x 10^6 – 2.2 x 10^7 cfu/ml. *Staphylococcus* was isolated from all the yoghurt samples analyzed (Nwagu and Amadi, 2010).

In a study to determine the microflora of some available yoghurt sold in Ibadan, Oyo state, South-Western Nigeria. The result revealed that yoghurt commercially produced in Ibadan were of high quality (Alli *et al.*, 2010)

In a study carried out in Abuja municipal area, Federal Capital Territory, Nigeria to determine microbial quality of yoghurt. The result showed that the mean total heterotrophic count was high. Fungal counts were equally greater than the safe limit. Coliform was absent in all the yoghurt samples while *Staphylococcus aureus* – a toxin producer was isolated in one of the yoghurt samples (Idise *et al.*, 2009)

In another study conducted in Minna, Niger state, North- Central Nigeria. The bacterial and fungal counts were higher than the standard allowed. Coliform was absent in 85% of the samples (Oyeleke, 2009)

However, in a study carried out in Khartoum, Sudan by Eissa *et al.* (2010). The mean bacterial count was 1.3 x 10^5 which is higher than the standard allowed. Coliform was detected while *Staphylococcus aureus* and *Salmonella* species were absent.

In a similar study carried out by El-Bakri and El-Zubeir (2009) in Khartoum state. Out of 144 samples, 43.75% of samples had coliform counts lower than 10^2cfu/ml. About 68.75% of samples had yeast and molds count lower than 10^3 cfu/ml. The study concluded that most of the yoghurt samples analyzed were of low quality.

In 2002, Younus *et al* investigated the quality of market yoghurt/Dahi obtained from local markets of Islamabad, Pakistan. The coliform count was nil or ignorable.

In other studies, yoghurt samples were obtained from local markets in Balikesir region, Turkey. In this research yeast and mold were detected in some of the yoghurt samples while Coliform and *E, coli* were found in low numbers in all the yoghurt samples (Irkin and Eren, 2008)

In a related study carried out in Middlesbrough (UK) to determine the quality of yoghurt. The result showed that the yoghurt is of high quality but an increase in the titratable acidity may affect the quality of the product (Istikhar *et al.*, 2010)

2.2 INGREDIENTS

In Japan new yoghurts containing antioxidant ingredients have been introduced. Milk of various mammals has been used for yoghurt production in various parts of the world. Human milk and animal milk from various countries, including Nigeria have been extensively studied (Ikem *et al.,* 2002)

Most of the industrialized yoghurt production uses cow's milk, whole milk, partially skimmed milk, skimmed milk or cream may be used. Nutritionally, enriched milk and its products with enhanced biological potential and without health risks are generally demanded. In order to ensure the development of the yoghurt culture, the following criteria for the raw milk must be met:

a. Low bacterial count

b. Free from antibiotics, sanitizing chemicals, mastitis and rancid milk

c. No contamination by bacteriophages.

To modify certain properties of the yoghurt, various ingredients may be added. To make yoghurt sweeter, sucrose (sugar) may be added at approximately 7%. For reduced calorie

7

yoghurt, artificial sweeteners such as aspartame or saccharin are used. Cream may be added to provide a smoother texture. The consistency and shelf stability of the yoghurt can be improved by the inclusion of stabilizers such as food starch, gelatin or locust-bean. These materials do not have significant impact on the final Flavor.

2.3.   STARTER CULTURE

Belgian scientists have examined the use of exopolysaccaride (EPS) producing strains of *Streptococcus thermophilus* as starter cultures in the production of dairy products such as yoghurt. Chinese workers have also studied the use of EPS producing yoghurt cultures with a view to improving the sensory and rheological properties of yoghurt. The starter culture for most yoghurt production is a symbiotic blend of *Streptococcus salivarius subsp. thermophilus* (ST) and *Lactobacillus delbrueckii subsp bulgaricus* (LB). Although, they can grow independently, the rate of acid production is much higher when used together than either of the two organisms grown individually (Sera *et al.,* 2009).

Streptococcus thermophilus grows faster and produces both acid and carbondioxide. The formate and carbondioxide produced stimulates *Lactobacillus bulgaricus* growth. On the other hand, the proteolytic activity of *Lactobacillus bulgaricus* produces stimulatory peptides and amino acids for use by *Streptococcus thermophilus*. These microorganisms are ultimately responsible for the formation of typical yoghurt flavor and texture (Sahan *et al.,* 2008). The yoghurt mixture coagulates during fermentation due to the drop in pH. The streptococci are responsible for the initial pH drop of the yoghurt mix to approximately 5.0. The lactobacilli are responsible for a further decrease to pH 4.0 (Lonner, 1993).

The following fermentation products contribute to flavor:

a.   Lactic acid

b.   Diacetyl

c.   Acetic acid

d.   Acetaldehyde: This is regarded as a major indicator of flavor in yoghurt. In a recent Turkish study on the effects of acetaldehyde levels in yoghurts, of the use of viscous and non viscous cultures revealed that highest acetaldehyde levels in yoghurt were obtained when non viscous starter cultures were used alone.

Reported is the launch of a new culture in Slovenia using *Bifidobacterium lactis* HNO19, produced by Danisco under the HOWARU name. Animal trials with this culture have demonstrated improved immune function response (Mann, 2005).

2.4   MANUFACTURING PROCESS

The general process of making yoghurt includes the following steps:

2.4.1   MODIFYING MILK COMPOSITION

The milk composition is modified before it is used to make yoghurt. This standardization process typically involves reducing the fat content and increasing the total solids. The fat content is reduced by using a standardizing clarifier and a separator (a device that relies upon centrifugation to separate fat from milk). The various ingredients are then blended together in a mix tank equipped with a powder funnel and an agitation system. In Egypt, recently scientists have reviewed developments in the manufacture and properties of the Egyptian yoghurt, Zabady.

2.4.2   YOGHURT PROCESSING STEPS

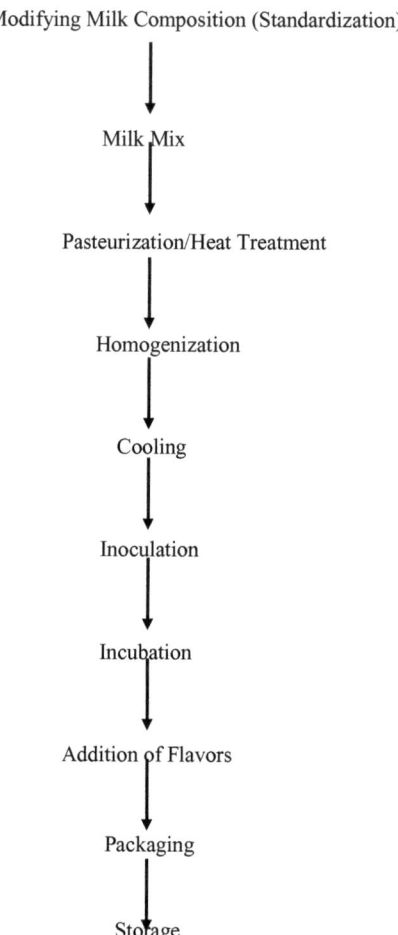

Modifying Milk Composition (Standardization)

Milk Mix

Pasteurization/Heat Treatment

Homogenization

Cooling

Inoculation

Incubation

Addition of Flavors

Packaging

Storage

2.4.3 PASTEURIZATION

The mixture is then pasteurized using a continuous plate heat exchanger for 30 minutes at 85°C or 10 minutes at 95°C. This step has many benefits.

a. First, it will destroy all the microorganisms in the milk that may interfere with the controlled fermentation process.

b. Second, it will denature the whey proteins in the milk.

c. Third, it will not greatly alter the flavor of the milk.

d. Finally, it helps release the compounds in milk that will stimulate the growth of the starter culture (Guarner *et al.*, 2005).

2.4.4 HOMOGENIZATION

The mix is then homogenized using high pressures of 200 - 250 bar. (20-25MPa) Homogenization is a process in which the fat globules in milk are broken up due to shearing forces into smaller, more consistently dispersed particles. This produces a much smoother and creamier end product. Homogenization has the benefit of giving a uniform product, which will not separate. Studies in the USA reported that yoghurt made from milk treated at 19.3MPa exhibited low yield stress, low water holding capacity and large clusters of coalesced micelles.

2.4.5. FERMENTATION

In the UK, the effect of temperature between 37°C and 46°C on the formation and rheology of yoghurt has been studied. After homogenization, the milk is cooled to 43 - 46°C. The yoghurt starter culture is added in a concentration of about 2%. A ratio of 1:1 *Streptococcus thermophilus* to *Lactobacillus bulgaricus*. It is held at this temperature for about 4 - 6 hours under quiescent (no agitation) conditions. This temperature is a

11

compromise between the optimums for the two microorganisms (*Streptococcus thermophilus* 39°C; *Lactobacillus bulgaricus* 45°C).

The acid level is found by taking a sample of the product and titrating it with sodium hydroxide. A value of 0.9% and a pH of about 4.4 are the current minimum standards. At this time, the jacket is replaced with cool water and agitation begins, both of which stop the fermentation (Bylund, 1995).

2.4.6. ADDITION OF FLAVORS

At the end of the fermentation, the coagulated product is cooled to 15-20°C. Fruits, flavors or coloring material may be incorporated. German dairy Onken has launched a new range of 1.5% functional yoghurts including flavors such as aloe vera, juniper, coffee and acerola mix.

2.4.7. PACKAGING AND STORAGE

The flavored or plain yoghurt is then packaged in small sizes to be stored at refrigeration temperatures (5°C) to slow down the physical, chemical and microbiological degradation (Bylund, 1995).

2.4.8. QUALITY CONTROL

Milk products such as yoghurt are subjected to a variety of safety testing. Some of these include tests for microbial quality, degree of pasteurization, and various forms of contaminants. In addition to safety tests, the final yoghurt product is also evaluated to ensure that it meets the specifications set such as pH, rheology, taste, color and odour. Guidelines for microbiological quality recommend that satisfactory yoghurt should not contain more than 10^8 cfu/g of starter.

< 1 coliform /g

< 1 mould /g

< 10 yeasts /g

2.5. YOGHURT PRODUCTS

There are two types of plain yoghurt:

a. Stirred style yoghurt

b. Set style yoghurt

In stirred style yoghurt, the yoghurt is incubated in tanks and cooled before packaging. In set style, the yoghurt is incubated and cooled in the package (Bylund, 1995).

2.5.1. VARIETIES

Scientists in the Ukraine have reported studies on the biotechnological parameters of yoghurt during which the probiotic properties of these products were confirmed in clinical trials. Manufacturers have responded to the growth in the yoghurt market by producing many different types of yoghurt, including low-fat and no-fat, creamy, drinking, bio-yoghurt, organic, baby, and frozen. Traditional yoghurt is thick and creamy. It is sold plain and in a wide assortment of flavors (Fadela *et al.*, 2009).

More unique flavors such as cream pie and chocolate have also been introduced. Cereals and nuts are sometimes added to yoghurts. Yoghurt makers also sell products with a varying level of fat. Low fat yoghurt, which contains between 0.5% and 4% fat, is currently the best selling. Diet no-fat yoghurt contains no fat at all. It also contains artificial sweeteners that provide sweetness while still reducing calories (Gadaga *et al.,* 2004).

Creamy yoghurt is extra thick, made with whole milk and added cream. Drinking yoghurt is a thinner product, which has a lower solids level than the typical yoghurt.

Bio-yoghurt is a different type of fermentation culture and is said to aid digestion.

Yoghurt that is made with milk from specially fed cows is called organic yoghurt. This type of yoghurt is claimed to be more nutritious than other yoghurts.

Baby yoghurt is specifically made for children. Recently, manufacturers have become quite creative in the types of yoghurt they produce using natural and artificial flavorings (Meyer, 2008).

2.5.2. PRESENTATION

a.   Dadiah or Dadih is a traditional West Sumatran yoghurt made from water buffalo milk. It is fermented in bamboo tubes.

b.   Dudh is a Sindhi-Curd, popular in India. People drink dudh along with food at intervals to help digestion and make food more delicious.

c.   Dahi is a yoghurt of the Indian subcontinent, known for its characteristic taste and consistency. It holds cultural symbolism, found in different flavors; sour yoghurt and sweet yoghurt. In India, it is often used in cosmetics mixed with turmeric and honey. Sour yoghurt is also used as a hair conditioner by women (Wisegeek, 2011).

d.   Labneh is a strained yoghurt used for sandwiches, popular in Arab countries.

e. Juju is the most famous type of Nepalese yoghurt where it is served as both an appetizer and dessert. It is used in local festivals, marriage ceremonies, parties, religious occasions, family gatherings and so on.

f. Ayran or Dhalla is a yoghurt-based, salty drink popular in Albania, Pakistan, Bulgaria, Turkey and Kazakhstan. It is made by mixing yoghurt with water and (sometimes) salt.

g. Some yoghurts often called "cream line", are made with whole milk which has not been homogenized, so the cream rises to the top.

h. In Europe, (including the UK and the US), sweetened, flavored yoghurt is the most popular type sold in single-serving plastic cups (Fabian, 2009).

2.6. SENSORY PROPERTIES

Croatian researchers have studied the aroma of probiotic yoghurts with and without supplements, with special reference to changes in aroma compounds occurring during storage as a function of time and temperature. Sensory properties refer to the aroma, taste, consistency and also to the viscosity and texture.

Yoghurt is distinguished by a typical and pleasant aroma attributable to the presence of sufficient quantities of acetaldehyde as a principal aroma compound. It relies upon a good balance of volatile fatty acids and is enhanced to some degree by the presence of diacetyl Yoghurt made from sheep's milk is characterized by a specific flavor and aroma, as distinct from yoghurt made from cow's milk. This concurs with work carried out by Fadela *et al.* (2009) in which sensory analysis revealed that the product made with ewe's milk was better compared to that made from skim milk. The milk acid and refreshing taste of yoghurt is attributable to the presence of lactic acid.

The aroma and taste combine to produce the flavor of yoghurt which is described as walnut-like. The flavor and aroma of yoghurt are affected by the properties of the cultures, method of propagation, quality of raw milk, processing treatments, methods of incubation and cooling (Bylund, 1995). Sensory properties are evaluated by organoleptic tests based on sensory perception which is connected with odour, taste and the sensation of flavor in the mouth.

2.7   FACTORS AFFECTING THE QUALITY OF YOGHURT

In the USA researchers studied the effects of different factors, including milk components, starters, manufacturing and handling processes on yoghurt texture. Numerous factors must be carefully considered during the manufacturing process in order to produce a high - quality yoghurt with the required flavor, aroma, viscosity, consistency, appearance, freedom from whey separation and long shelf life. These factors include:

a.   Choice of milk

b.   Milk standardization

c.   Milk additives

d.   Homogenization

e.   Heat treatment

f.   Choice of culture and preparation

2.7.1. CHOICE OF MILK

The milk used must be free from anti microbials such as antibiotic residues and sanitizers used for material cleaning. This is to avoid the inhibition of the starter culture which might result not only in economic loss but may allow pathogens to grow (IDF, 2002)

2.7.2. MILK STANDARDIZATION

A recent study in Australia reported that standardization of total solids content with dried skim milk was not sufficient to produce yoghurts with consistent physical characteristics throughout a season. Another group of Australian scientists studied the effects of fortification of milk with 2% of either whey powder (WP), skim milk powder (SMP) or whey protein concentrate (WPC) on composition, pH, firmness, viscosity, syneresis and microstructure of yoghurt. The results indicated that WP and SMP supplementation reduced viscosity and firmness, while WPC supplementation increased these values.

Yoghurt may have a fat content of 0 -10%. A fat content of 0.5-3.5% is best (IDF, 2002)

2.7.3. MILK ADDITIVES

Sweeteners and stabilizers may be used as additives in yoghurt production (IDF, 2002).

2.7.4 . HOMOGENIZATION

Homogenization is employed to prevent creaming during the incubation period and to assure uniform distribution of the milk fat.

Homogenization also improves the stability and consistency. The milk is homogenized at 20-25 MPa and 60-70°C to obtain optimum physical properties (IDF, 2002).

2.7.5. HEAT TREATMENT

A study in Spain reported that manothermosonication treatment (MTS), which is the simultaneous application of heat and ultrasound under moderate pressure could result in yoghurts with rheological properties superior to those found in control yoghurts made from untreated milk..The milk is heat-treated to about 80-90°C for about 30 minutes. This is to clear all the microflora in the milk so that the starter will encounter little competition.

The heat process also improves the milk as a growth medium for the starter by inactivating immunoglobulins, expulsion of oxygen to produce microaerophilic environment and through the release of sulfhydryl groups as stimulatory agents (Sahan *et al.*, 2008).

2.7.6. CHOICE OF CULTURE AND PREPARATION

The handling of the starter for production of yoghurt demands maximum precision and hygiene. Culture laboratories now use advanced techniques to produce customized yoghurt culture to satisfy specific flavor and viscosity requirements (Wisegeek., 2011).

2.8. DEFECTS OF YOGHURT

Considerable deviations in the organoleptic properties of yoghurt usually indicate some defects. The sensory defects of yoghurt may be divided into three groups. They are:

a.   Defects resulting in changes of appearance; settled, fermented, unclean, aged and formation of colonies.

b.   Defects of flavor and aroma; high acid, bitter, burnt, yeasty, stale, mealy, cheesy, rancid and;

c.   Defects of consistency; whey separation, soupy, splitted and viscosity, slimy, tough, liquid and phase separation.

2.8.1. BACTERIOPHAGE ATTACK OF YOGHURT

Researchers in Argentina have isolated three bacteriophage-sensitive strains of *L. delbrueckii* subspecies *bulgaricus* from commercial yoghurt starters and their indigenous bacteriophages have been characterized with regard to the phage absorption process. Bacteriophages are viruses, i.e. bacterial parasites. A bacteriophage problem normally occurs in yoghurt over time, resulting in longer fermentation time, lack of flavor etc. The most common bacteriophage problems are normally observed as:

a.   Delay in fermentation time

b.   Lack of viscosity

c.   Lack of flavor

d.   Slimy structure

e.   Grainy structure of product

2.8.2. YOGHURT MICROFLORA

The different organisms found in yoghurt may be divided into three groups, namely:

• Essential microflora

• Non-essential microflora; and

• Contaminants

i.   Essential Microflora:

The following organisms should be regarded as the essential microflora of yoghurt:

- *Streptococcus thermophilus*
- *Lactobacillus bulgaricus*

ii.   Non-Essential Microflora:
This group includes:

a.   Hetero fermentative lactic acid bacteria.

b.   Some other homo fermentative lactic acid bacteria other than ST and LB.

iii.   Contaminants:

These organisms are entirely undesirable since they substantially decrease the organoleptic and hygienic properties of yoghurt. They include:

a. Yeasts and moulds
b. Coliform bacteria
c. Other undesirable organisms

2.8.3. PACKAGING

In the work carried out by Saint-Eve *et al,* 2008. It was reported that to preserve the inherent quality of yoghurt during storage and in particular, its physicochemical and sensory characteristics, packaging is essential.

Yoghurt is perishable, so a clean, non-tainting package is absolutely essential. The package should also protect the product from mechanical shock, light and oxygen.

Packs can be made of glass, plastic or paperboard. The fundamental functions of packaging are:

a.   To reduce food spoilage and waste

b.   To protect nutrients and flavor

c.   To enable efficient food distribution

d.   To maintain product hygiene

e.   To increase food availability

f.   To convey product information

2.8.4. LONG LIFE YOGHURT

In view of the tendency towards larger and more centralized production units, the markets are becoming geographically larger and transport distances are becoming longer. In certain cases, only one delivery every week is economically justifiable. This in turn, necessitates extending the shelf life of the product beyond the normal value. This can be achieved by:

a.   Heat treatment of the finished product, either in the pack or immediately before packaging. A stabilizer must then be added before treatment.

b.   Production and packaging being carried out under aseptic conditions.

It is hoped that in the near future, the types of organisms in the cultures will be varied so that yoghurt is produced much faster and lasts longer than conventional yoghurt (Fadela *et al.*, 2009).

CHAPTER THREE

3.0 MATERIALS AND METHODS

3.1 SAMPLING PROCEDURE

3.1.1 SAMPLING SITES

Samples were obtained from retail outlets; super markets, stores and vendors in Kano Metropolis.

3.1.2 SAMPLE COLLECTION

Samples were purchased on weekly basis and transported to the Microbiology laboratory of Bayero University in an ice box for six months (between June 2010 - November, 2010)

One hundred samples (10 samples from each manufacturer) of commercially produced yoghurt were obtained and analyzed.

3.1.3 DESIGN

The samples which represent ten different manufacturers (YB_1, YB_2 – YB_{1O}) had ten samples each from each different manufacturer and were analyzed for physicochemical, sensory and microbiological qualities. Manufacturing details such as: NAFDAC number, batch number, manufacturing date, expiry date, ingredients as well as content were monitored.

3.2 PHYSICO-CHEMICAL AND SENSORY ANALYSIS OF YOGHURT SAMPLES.

3.2.1 MEASUREMENT OF TEMPERATURE

The temperature was measured by dipping the glass thermometer into the yoghurt sample and the corresponding reading taken (AOAC, 2005).

3.2.2 MEASUREMENT OF VISCOSITY

Viscosity was measured using a viscometer model DV-E viscometer. The yoghurt samples were warmed to 20°C and poured into 250ml container.

The spindle core (nozzle) was then attached to the viscometer with the aid of the screw and tightened up with the guard legs in place.

After making sure the marker was on zero, it was then set on for 3 minutes until a stable mark was reached. The value was then read and expressed in Centipoise (AOAC, 2005).

3.2.3 DETERMINATION OF SPECIFIC GRAVITY

The yoghurt sample was pre-warmed to 40°C and then cooled to 20°C. The sample was poured gently into the lacto-densitometre jar and the lacto-densitometre was allowed to gently slide in the yoghurt.

The jar was topped up and the reading on the lactometer to the nearest 0.1 reading after it reached equilibrium was taken and expressed in g/ml (AOAC, 2005).

3.2.4. DETERMINATION OF pH

Struers buffers of pH 4 and pH 7 were used. The buffers were standardized by using the knob "cal" to adjust their temperature to 25°C.

The needle was first washed with distilled water and cleaned with cotton wool before taking the pH of each sample.

The glass electrode was pushed into the sample to about 3/4 of the sample, then swirled for some seconds and allowed to become steady before taking reading on the Labtech pH metre (AOAC, 2005),

3.2.5 DETERMINATION OF TITRATABLE ACIDITY

Ten (10) ml of yoghurt sample was pipette into a conical flask to which 1 ml of 0.5% phenolphthalein indicator was added.

The content was mixed by gentle shaking and 1 ml of 0.1N Sodium hydroxide was added rapidly from the burette. This was followed by drop mix addition of the 0.1N NaOH till a faint pink color, which persisted for 15 seconds appeared (AOAC, 2005).

3.2.6 ESTIMATION OF FAT CONTENT

Ten milliliter sulphuric acid, was measured in to the milk butyrometre. About 5 ml of the sample warmed to $20^{\circ}C$ and 5 ml of distilled water were poured successively into the butyrometer. The neck of the butyrometer was firmly closed with the stopper.

The butyrometer was shaken in a stand and inverted twice during the process. The butyrometer was centrifuged after mixing for a speed of 1100 r.p.m. The scale reading corresponding to the lowest part of the fat meniscus and to the surface of separation of the fat and the acid was read. (AOAC, 2005).

3.2.7. SENSORY EVALUATION

Sensory quality of yoghurt products was evaluated by a jury of 5 panelists with a 9-point Hedonic scale and involved the .following parameters: color, smell, flavor and general acceptability.

The scale and categories were as follows:

Liked extremely   =   9

Liked very much   =   8

Liked moderately   =   7

Liked slightly   =   6

Neither liked or disliked =   5

Disliked slightly   =   4

Disliked moderately   =   3

Disliked very much   =   2

Disliked extremely   =   1

(Kroll *et al*, 1998)

A strict protocol was imposed to panelists to minimize variability at each session.

Subjects tasted samples and were asked to keep the yoghurt in the mouth for 12 seconds before scoring. The yoghurt samples were presented in random order. Water was used for rinsing mouth between samples. A small period of several minutes was required between tasting samples. (IDF, 2002).

3.3. MICROBIOLOGICAL ANALYSIS OF YOGHURT SAMPLES

3.3.1. SAMPLE PREPARATION

For each yoghurt sample, twenty-five gram (25g) was weighed into a sterile wide mouth container, 225ml of sterile buffered distilled water was added and shaken until a homogenous dispersion was attained.

One milliliter (1.0ml) was pipetted into a tube containing 9ml of the Buffered Peptone

Water (BPW). This was mixed carefully 10 times with a pipette.

From the first dilution, 1ml was transferred with the same pipette to second dilution tube containing 9ml of the BPW. A fresh pipette was then used to mix and repeat using 3rd, 4th and 5th tube (FAO, 1993)

3.3.2. ENUMERATION OF BACTERIA AND FUNGI

3.3.2.1. ENUMERATION OF MESOPHILIC AEROBIC BACTERIA

One milliliter (1.0ml) of the food homogenate and of each dilution of the homogenate was pipetted into each of the appropriately marked duplicate dishes. 20ml of Nutrient Agar (NA) was poured into each Petri dish kept at $45 \pm 1°C$ within 15 minutes of the time of original dilution.

The sample dilution and agar medium were mixed thoroughly and uniformly, and allowed to solidify. The prepared dishes were inverted and incubated at $30 \pm 1°C$ for $48 \pm 2h$. Following incubation, all colonies on dishes containing 30-300 colonies were counted and the results recorded per dilution counted. (IDF,2002).

3.3.2.2 ENUMERATION OF COLIFORMS USING MOST PROBABLE NUMBER (MPN)

One milliliter of each of the decimal dilutions of the yoghurt homogenate was inoculated into each of three separate tubes of Lauryl Sulphate Tryptose (LST) broth (containing inverted Durhan tubes). The LST tubes were incubated at $37°C \pm 1°C$ for 24 and 48 hours. Tubes showing gas production after 24 hours were recorded and negative tubes were incubated for further 24 hours. Tubes showing gas production were again recorded.

For confirmatory test, the gas positive LST tubes were then sub-cultured in brilliant green lactose bile broth (BGLB). The BGLB tubes were then incubated at 37°C $\pm$ 1°C for 48 hours. The formation of gas confirmed the presence of coliform bacteria (FAO, 1993).

3.3.2.3 ENUMERATION OF YEASTS AND MOULDS

One milliliter of each dilution was pipetted into each of the appropriately marked duplicate Petri dishes. About 20 milliliter of malt extract agar with antibacterial agent incorporated tempered to 45°C was poured into each Petri dish.

After solidification, plates were inverted and incubated at 20-25°C for five days. Where fast growth occurred, colonies were counted first after 3 days and the slow growth were again counted after the fifth day. Colonies were counted and reported as yeast and mould count per ml (IDF, 2002).

3.3.3 ISOLATION AND CHARACTERISATION OF BACTERIA AND FUNGI

3.3.3.1 ISOLATION OF *ESCHERICHIA COLI.*

A loopful transfer was made from each gas positive tube of LST to a separate tube of EC broth. The EC tubes were incubated for 48 hours at 44.5°C, production of gas indicated positive. From each positive tube, one plate Levine's Eosin Methylene Blue (L-EMB) agar was streaked in a way to obtain discreet colonies and were incubated for 18-24 hours at 35°C.

Transfers of the culture were made from each L-EMB plate to nutrient agar slants and incubated for 24 hours at 35°C. Gram stain of each culture and biochemical tests were carried out (FAO, 1993).

3.3.3.2. ISOLATION OF *STAPHYLOCOCCUS AUREUS*

This was carried out according to the method described by the International Diary Federation (IDF, 2002). The 0.25ml of yoghurt homogenate and dilutions were inoculated on the surface of previously dried Baired-Parker (Difco) agar plates incorporated with egg yolk and spread with a sterile bent glass rod. Duplicate plates were prepared from each dilution.

The plates were inverted and incubated at 37°C for 24 and 48 hours. The colonies on Baired-Parker agar were colourless and so the characterization could not continue from there.

3.3.3.3 ISOLATION OF *SALMONELLA* SPECIES

This was carried out according to the method described by (IDF, 2002). The food homogenate (25g yoghurt sample mixed with 225ml BPW) was transferred to a sterile 500ml conical flask and incubated at 37°C for 24 hours. About 10 milliliter of the pre - enrichment was transferred to 100ml of Selenite medium and incubated at 44°C for 48 hours.

Streak was made from the enrichment medium on previously dried Deoxycholate Citrate Agar (DCA) and incubated at 37°C for 24 hours.

The colors appeared as lactose fermenters and there for were not supposed to be *Salmonellae.*

3.3.3.4 MICROSCOPY FOR YEASTS AND MOULDS

This was carried out according to the method of Cheesbrough (2000). A drop of distilled water on clean grease-free slide, a tiny portion of the mould was placed in it and teased with clean needle. It was covered with cover slip and examined under the microscope starting with a low power objective, then the high power (40x) objective for a better field view magnification.

3.4. GRAM STAIN PROCEDURE

This was carried out according to the method of Cheesebrough (2000). The smear was prepared by placing a drop of sterile water in the middle of a clean slide.

An inoculating loop was sterilized by flaming, cooled, touched the bacterial colony and rubbed in the drop of water which was spread into a thin smear.

The smear was air-dried and the reverse side of the slide was passed quickly 3 times over a flame to fix the bacteria. The fixed smear was covered with crystal violet stain for 30 seconds. The stain was washed off rapidly and covered with Lugol's iodine for 30 seconds.

The iodine was washed off with clean water and decolorized with ethanol which was also rinsed with water. The smear was covered with safranin for 2 minutes which was-washed off with clean water. The back of the slide was wiped clean and placed in a draining rack.

The smear was then examined microscopically, first with the 40x objective and then with the oil immersion objective (100x).

3.5   BIOCHEMICAL TESTS

All stored Gram negative isolates were biochemically characterized by using Indole, Methyl-Red, Voges-Proskauer and Citrate test.

3.5.1 INDOLE TEST

This was carried out according to the method of Cheesebrough (2000). The test organism was inoculated in a bijou bottle containing 3ml of sterile tryptone water and incubated at 37°C for 48 hours. About 0.5ml of Kovac's reagent was added. It was gently shaken and examined for red color in the surface layer within 10 minutes

.

3.5.2 METHYL- RED TEST

This was carried out according to the method of Cheesebrough (2000). The test organism was inoculated in a tube of Voges Proskauer (MR-VP) medium and incubated for 48 hours at 35°C. After which 5 drops of methyl red indicator was added and observed for red color

3.5.3 VOGES-PROSKAUER TEST

This was carried out according to the method of Cheesebrough (2000). After completion of the MR test, 0.6ml of 5% alpha-naphtol solution was added and 0.2ml of 40% KOH solution. The tube was shaken, sloped and examined for eosin-pink coloration after 15 minutes.

3.5.4 CITRATE UTILIZATION TEST

This was carried out according to the method of Cheesebrough (2000). Slopes of Simmon's citrate agar were prepared in bijou bottles and stored at 2-8°C. Using a sterile straight wire, the slope was first streaked with a saline suspension of the test organism. The butt was stabbed and incubated at 35°C for 48 hours and examined for bright blue colour in the medium.

3.5.5 MOTILITY TEST

This was carried out according to the method of Cheesebrough (2000). The test organism was inoculated in a tube containing 10ml peptone water for 24 hours at 37°C. A drop was placed on the slide with cover slip and examined for motile cells.

3.6 STATISTICAL ANALYSIS

The data obtained from the study were analyzed by simple analysis of variance (ANOVA), using single factor at $P>0.05$ to provide the possibility of comparing the levels of contamination of different yoghurt brands.

CHAPTER FOUR

4.0. RESULTS

The results of the physicochemical and sensory qualities of the yoghurt samples are presented in Table 4.1. The highest temperature was 28°C, while the lowest was 0°C. The viscosity ranged between 8.9 - 65.2 centipoise. The specific gravity of the yoghurt samples ranged between 1.042 and 1.063. YB_9 had the lowest specific gravity, while YB_{10} had the highest specific gravity. The pH of all the yoghurt samples analyzed fell within the range of 1.44 and 5.35. The values for titratable acidity were in the range of 4 (0.38) to 12.8 (1.22) % lactic acid.

The fat content ranged from 1.5 to 2.5 with the fat contents of YB_5 significantly higher than the fat content of other brands of yoghurt samples analyzed.

Sensory evaluation was carried out using a 9 - point hedonic scale with YB_5 having a significantly higher overall acceptability ($p<0.05$).

The aerobic mesophilic bacterial counts ranged between 1.9×10^2 and 2.2×10^5 cfu/ml, a significant difference ($p<0.05$) in bacterial count was observed between the samples with YB_4 and YB_7 having the highest scores. There was significant difference ($p<0.05$) in coliform count between the samples with YB_4 having the highest count. fungal counts ranged from 8.6×10^1 - 2.5×10^4. A significant difference ($p<0.05$) in fungal counts was observed between the samples with YB_6 having the highest count. *Escherichia coli* was positive in 17(34) % of the yoghurt samples analysed. None of the samples for detection of *Staphylococcus sp.* and *Salmonella sp.* in all the yoghurt brands analysed was positive.

Table 4.1. Mean physico-chemical and sensory qualities of yoghurt brands sold in Kano Metropolis

Sample	Temp (°C)	Analysis Fat Content (%)	Viscosity (Centipoise)	Specific Gravity (g/mls)	pH	Acidity (%)	Sensory evaluation Color	Flavor	Smell	General Acceptability
YB1	0	1.80	49.80	1.047	3.06	9.00	8	8	7	7
YB2	O	1.60	8.90	1.044	3.03	9.40	5	7	4	6
YB3	27	1.50	11.40	1.045	3.05	9.20	7	7	4	6
YB4	26	1.50	12.80	1.043	3.44	5.20	6	7	7	6
YB5	19	2.50	35.40	1.046	5.35	4.00	7	9	7	9
YB6	28	2.20	60.10	1.054	1.55	11.00	6	5	7	6
YB7	14	2.30	57.90	1.049	4.30	6.50	6	6	7	8
YB8	24	1.80	28.80	1.046	4.34	8.00	7	7	7	7
YB9	26	1.50	9.54	1.042	4.04	10.50	8	8	7	8
YB10	28	2.40	65.20	1.063	1.40	12.80	6	5	7	6

KEY: YB = Yoghurt Brand,

1- Disliked extremely, 2- Disliked very much, 3 - Disliked moderately, 4 - Disliked slightly, 5 - Neither liked or disliked, 6 - Liked slightly, 7 - Liked moderately, 8 - Liked very much, 9 - Liked extremely.

33

Table 4.2 Microbiological qualities of yoghurt effect for sale in Kano Metropolis

Sample	Mean-bacterial count (cfu/ml)	Coliform-count (MPN/ml)	E.coli. No (%) contaminated	Salmonella	S. aureus	Fungal (cfu/ml)	counts
YB$_1$	6.4×10^2	480	1(2)	-	-	4.2×10^3	
YB$_2$	8.8×10^3	485	1(2)	-	-	5.8×10^2	
YB$_3$	1.1×10^5	973	3(6)	-	-	8.7×10^2	
YB$_4$	2.2×10^5	980	3(6)	-	-	2.1×10^3	
YB$_5$	1.9×10^2	97	1(2)	-	-	8.6×10^1	
YB$_6$	2.1×10^5	966	2(4)	-	-	2.5×10^4	
YB$_7$	2.2×10^5	965	2(4)	-	-	5.2×10^2	
YB$_8$	5.8×10^2	9	1(2)	-	-	4.6×10^2	
YB$_9$	1.2×10^3	481	1(2)	-	-	8.0×10^2	
YB$_{10}$	1.5×10^3	5	1(2)	-	-	1.2×10^3	

KEY: YB = Yoghurt brand, (-) = Absent, (+) = Present.   Each value is a mean of 10 determinations.

Table 4.3:   Cultural, Morphological and Biochemical characteristics of the Isolates

No of samples contaminated	Cultural characteristics	Morphological	Indole	Methyl Red	-Voges Proskauer	Citrate	Motility	Confirmed isolates
17	Aerobic and facultative anaerobe. Ferments lactose, producing smooth pink colonies on L - EMB agar.	Gram negative rods	+	+	-	-	+	*E.coli*
0	Aerobic, optimum growth temperature is 35-37°C. they appear as small black colonies on Baird Parker agar.	Gram positive cocci,						-
0	Aerobic and facultative anaerobes. appear as black colonies on Deoxychollate Citrate Agar (D.C.A)	Gram negative rods, non Salmonellae– sporing, non - capsulate						-
32	Colonies appear white to beige or grey and fast. growing. They lack rhizoids	They grow well at 20-25°c found on plant, food and rotten matter.						*Mucor* species

KEY :   (-) = Negative, (+) = Positive

CHAPTER FIVE

5.0.   DISCUSSION, CONCLUSION AND RECOMMENDATION

5.1.   DISCUSSION

The physical characteristics such as temperature, viscosity, specific gravity, pH, titratable acidity and fat content are important parameters in studying the physicochemical compositions and nutritional aspects of yoghurt.

Actual temperature of storage in markets is important for bacteria viability in yoghurts. Industrial standards recommends for yoghurts, a holding temperature not higher than 8°C. In this study, the mean temperature was 19.3°C which was higher than the findings of Egwaikhide and Faremi, 2010. The data suggests that warmer weather and bad storage conditions are the principal causes of higher levels of contamination.

Also, a factor which could not be excluded or mis-evaluated is the inconsistency of power supply that subjects the product to a high temperature, shortening its shelf life and increasing the acidic flavor.

In this study the viscosity ranged from 8.9 - 65.2cp, which was lower than that reported by Okoye and Animalu (2009) but comparable to that reported by Fadela *et al.* (2009). The variation in viscosity could be attributed to the Ph, temperature, stabilizer or the dry matter content used by individual manufacturers.

The specific gravity of the yoghurt samples collected in this study ranged from 1.042-1.063 which is within the International standards which is 1.040 – 1.070. The specific gravity is mainly due to the water contents present in the sample.

The pH is the parameter that determines the sample acidity and alkalinity. The pH range found in the current study was comparable with those reported by Okoye and Animalu (2009), Egwaikhide and Faremi (2010), but lower than those reported by Eissa *et al.* (2010). The current minimum pH is 4.4 the standard given by FAO, (1979).

Frazier and Westhoff (1978) reported that foods with ultimate pH of below 5.6 are more susceptible to habitation by acidophilic bacteria. This may make the yoghurt samples susceptible to habitation by acid tolerant bacteria, especially the *Lactobacillus species.*

Sokolinska *et al.* (2004), indicated that the pH values of milk decreased during the manufacturing process, from the time it was inoculated with bacterial cultures to the time when it was manufactured ranging from 6.7 - 4.3. Moreover, according to El-Bakri and El-Zubeir (2009) lactic strains have the ability to ferment lactose into lactic acid, with an increase in acidity and a decrease in pH of fermented product. The pH of yoghurt should not allow for bacterial spoilage and will instead, select for growth of yeasts and moulds.

The results of the titratable acidity obtained in this study agreed with the findings of Okoye and Animalu (2009), Eissa *et al.* (2010). However, it was lower than that reported by Egwaikhide and Faremi (2010). Standard range of acidity is (0.14 - 0.16). The microflora applied in the technology of yoghurt production most commonly is made up of strains of *Lactobacillus delbrueckii Subspecie bulgaricus* and *Streptococcus thermophilus.* These bacteria utilize milk lactose in the course of the fermentation with different speed and in different directions in the acidity of the final product (Valera-Moleiras *et al.,* 1992).

The mean value of fat content is 1.9. which was below that reported by Eissa *et al.* (2010), Egwaikhide and Faremi, (2010) for yoghurt brands in Kaduna Metropolis but similar to that of Shojaei and Yadollahi (2010). The decrease in the fats content of these brands of the yoghurt may be as a result of aerobic mesophilic bacteria utilizing lipids for the synthesis of cell membrane and other cellular organelles in order to increase their population. Vinderola *et al.* (2008) reported that high fat content yoghurts are more inhibitory for probiotic bacteria. According to FAO standard, fat content of 0.5 - 10 is good but fat content of 3.0 is the best.

The sensory quality of the yoghurt samples was determined using the 9-point Hedonic Scale with YB_5 scored significantly higher than the other brands in terms of general acceptability ($p<0.05$). This could be due to good manufacturing practice and packaging on the part of the manufacturer.

The aerobic mesophilic bacterial count ranged from $1.9 \times 10^2 - 2.2 \times 10^5$ cfu/ml which were higher than the value obtained by Okoye and Animalu (2009) as well as Egwaikhide and Faremi (2010). The bacterial count should not exceed 5.0×10^4 cfu/ml, the standard given by FAO, 1979.This difference may be due to non septic handling and inadequate heat treatment during preparation process. .

In the study carried out by: Younus *et al.* (2002), coliform was not detected, in any of the yoghurt samples analyzed so also the works of Birollo *et al.* (2001), Okoye and Animalu (2009), Ewaikhide and Faremi (2010). However, in this study coliform was observed in some yoghurt samples but the result obtained was higher than that reported by Irkin and Eren (2008). Based on standards of pasteurized milk, the coliform bacteria count must not exceed 5 cfu/ml. Presence of coliforms in pasteurized milk might be from poor hygiene of pasteurization and packaging process.

In this research *Escherichia coli*, an indicator organism was isolated from some yoghurt samples which concurs with previous reports (Eissa *et al.*, 2010) but was not detected in Irkin and Eren, study. This might be attributed to low level of hygiene during processing of yoghurt. *Escherichia coli* is commonly used as surrogate indicator and must be negative to be safe for consumers based on European union regulations. Its presence in food generally indicates direct or indirect fecal contamination. Substantial number of *Escherichia coli* in food suggests a general lack of cleanliness in handling and improper storage.

In this study, *Staphylococcus aureus a* toxin producer was not detected in any sample. Results obtained are in agreement with the results of Egwaikhide and Faremi (2010). This could be due to proper pasteurization and post pasteurization practices of manufacturers. The sample was considered safe when the count did not exceed 10cfu/ml, FAO, (1979)

In the study by Eissa *et al.* (2010), *Salmonella* species was not detected in any of the yoghurt samples. This is in alliance with the current research. This could be due to addition of inhibitory substances to prevent growth of pathogens and to prevent re-use of starter cultures of choice. The sample was considered safe when *Salmonella* is not detected in the food, FAO,1979 .

The study by Egwaikhide and Faremi (2010), yeast and mould were not detected in any of the samples, so also the work of Birollo *et al.* (2001). However in this research yeast

39

and mould were detected in some of the yoghurt samples. This is in alliance with the results of Irkin and Eren (2008). The presence of the organisms as contaminants may be used as indices of sanitary conditions. Their presence in yoghurt can swell packages and limit shelf life when levels reach 10^5 - 10^6 cfu/ml. To pass the test of safety the mould count should not exceed 1 cfu/ml.

5.2. CONCLUSION

In the present study, preliminary investigation was carried out to ascertain the physicochemical, sensory and microbiological qualities of various yoghurt brands marketed in Kano metropolis. The results show that two of the pathogenic strains *Staphylococcus aureus* and *Salmonella species* were not detected in any of the samples analyzed. This could be due to the pasteurization temperature.

This study concludes that the yoghurt samples have good chemical quality when compared to international standards. However, the microbiological quality was lower than the international standard ($p < 0.05$) in some samples, although, not to the level that can compromise the health of the consumer. However, proper hygiene and storage ethics should be maintained due to the prescence of microorganisms in some of the products. These findings may be useful to the concerned governmental parties and health agencies to monitor the quality of yoghurt products in the market.

5.3. RECOMMENDATIONS

a. Producers must pay more attention to the hygienic practices in their production facilities to improve the microbial quality standards.

b. Licenses given for small yoghurt producers must be issued after the assurance of a minimum level of good manufacturing practice.

c. Manufacturers producing ultra high temperature (UHT) yoghurts with longer shelf life should have a mechanism of recalling expired products that are still in the market.

d. Yoghurt is a perishable product hence it should be stored in refrigerators with good light supply.

e. Products should be moved from factories to depots/distributors in insulated trucks equipped with functional thermokings.

f. Production staff should go for routine medical examination to check for infection of pathogenic organisms.

g. Manufacturers should ensure that products carry batch number and must be legible enough.

REFERENCES

Alli, J.A. Oluwadan, A. Okonkwo, I.O. Fagade, E. Kolade, A.F. Ogunleye, V.O (2010) Microbial assessment and Microbiological quality of some commercially prepared yoghurt retailed in Ibadan, Oyo state, South-Western Nigeria. *British Journal of Diary Science 1 (2):34-38*

AOAC, (2005). Association of Official Analytical Chemists: *Official Methods of Analysis* of AOAC International. 18th Edition, Gaithersburg, MD, United States of America.8: 8-25

Beel, B (1994). The hidden world of yoghurt. *View magazine.* Pp 6 – 19.

Birollo, G.A.; Reinheimer, G.A & Vinderolla, C.G. (2001). Viability of lactic acid microflora in different types of yoghurt, *Food Research International,* 33: 799-805.

Bylund, G. (1995). *Diary Processing Handbook.* Tetrapak Processing Systems. A.B. Lund, Sweden. Pp 213-262

Cheesbrough, M.(2000). .*District Laboratory Practice in Tropical Countries,* Part 2. Cambridge university press. United Kingdom. Pp 38-184

Columbia Electronic Encyclopedia, (2003). *Encyclopaedia of food and culture* Fermented milk:.6th edition. Columbia University Press. Pp 1-2

Courtin, P. and Rul, F. (2004). Interactions between microorganisms in a simple ecosystem: Yoghurt bacteria as a study model. *Lait, 84*: 125-134

Dalgic, A.C. and Belibagh, K.B. (2008). "Hazard analysis critical control points implementation in traditional foods: a case study of Tarhana processing". *International Journal of Food Science and Technology,* 43:1352-1360

Egwaikhide, P.A. and Faremi, A.Y. (2010). Bacteriological analysis of locally manufactured yoghurt, *Electronic Journal of Environmental, Agricultural and Food Chemistry,* 9(11):1679-1685

Eissa, E.A, Mohammed, A, Yaqoub, E.A and Babiker, E.E, (2010). Physicochemical, microbiological and sensory characteristics of yoghurt produced from Goat milk. *Electronic journal of Environmental, Agricultural and Food Chemistry.* 22: (8) 210-215

El-Bakri, J.M and El-Zubeir, I.E. (2009). Chemical and microbiological evaluation of plain and fruit yoghurt in Sudan. *International Journal of Dairy Science,* 4:1-7.

Fabian, E (2009) Influence of daily consumption of probiotic and conventional yoghurt in the plasma lipid profile in young healthy women, *Annals of Nutrition and Metabolism.* 50(4): 387-393

Fadela, C; Abderrahim, C. and Ahmed, B. (2009). "Physicochemical and rheological properties of yoghurt manufactured with Ewe's milk and skim milk".*African Journal of Biotechnology,* 8 (9): 1938-1942.

FAO (1993). Food and Agricultural Organisation. Joint FAO/WHO committee of Government experts on the Code of principles concerning milk and milk products, *Procedural Manual* Rome, Italy. 12th edition. pp 1020-1040

Food and Agricultural Organization (1979*). Manual of Food Quality Control, 4. Microbiological Analysis, FAO food and nutrition paper,* FAO Rome, Italy paper 14(4)

Frazier, W.C and Westhoff, D.C (1978) Food as a substrate for microorgarnisms. *Food Microbiology* 5th edition. Mc Graw-Hill publishing company limited, New York,540-545

Gadaga, T.H.; Nyanga, L.K. and Mutukumira, M.N. (2004). The occurrence, growth and control of pathogens in African fermented foods. *African Journal of Food Agriculture, Nutrition and Development,* 4(1) 25-32.

Guarner, F. G. Perdigon, G. Corthier, S. Salminen, B. Koletzko, A and Morelli. L (2005). Should yoghurt cultures be considered probiotic?. *British Journal of Nutrition.* 93: 783-786.

http//: www.wisegeek.com / what - are – the – different – types – of – yoghurt. html

Idise, O.E, Ameh, J.B, Ado, S.A, and Isukuru, E.F,.(2009) Microbial quality of nono obtained from Abuja municipal area, Nigeria. *Biological and Environmental Sciences Journal for the Tropics* 6(1): 40 – 44.

Ikem, A, Nwankwoala, A, Odueyungbo, S. Nyavor, K, and Egiebor, N. (2002).Levels of 26 elements in infant formular from U.S.A, U.K and Nigeria by microwave digestion and ICP-OES. *Food Chemistry* 77(4):439-447.

IDF (2002) International Dairy Federation. Fermented milk: Proceedings of the IDF seminar on aroma and texture of fermented milk, held in Kolding, Denmark. 301: 280-315.

Irkin, R. and Eren, U.V (2008). A research about viable *Lactobacillus bulgaricus* and *Streptococcus thermophilus* numbers in the market yoghurts. *World Journal of Dairy and Food Sciences,* 3(1):25-28.

Istikhar, H. Ahiq, R. and Nigel, A. (2010) Quality comparison of probiotic and natural yoghurt. *Science Alert* 12: 23 - 27

Kerr, T.J and McHale, B.B (2001*) Applications in General Microbiology: A laboratory manual.* 6th edition, Hunter textbooks incorporated. Winston-salem. Pp 221-230

Mann, E. (2005). Yoghurts Part One; the first part of Dr Ernest Mann's review of yoghurt literature, Dairy World, fermented technology, *Diary Industries International* 25: 37-38

Meyer, A.L(2008). Daily intake of probiotics as well as conventional yoghurt has a stimulating effect on cellular immunity in young healthy women. *Annals of Nutrition and Metabolism.* 50 (3): 282-289

Lonner, C. (1993)"Living Lactic acid – bacteria - vaccine of the future" *Scandinavian Journal of Nutrition / Naringsforksning.* 37 :132-137.

Mayo, B. (1993). The proteolytic system of lactic acid bacteria. *Microbiologia Madrid Spain,* 9 (2): 90 - 106.

Nwagu, T. N. and Amadi, E.C (2010) Bacteria population of some commercially prepared yoghurt sold in Enugu state, Eastern Nigeria. *African Journal of Microbiological Research*: 4 (10): 984-988

Okoye, J.I. and Animalu, I.L. (2009) Evaluation of physicochemical and microbiological properties of stirred yoghurt stabilized with sweet potato (*Ipomea Batata) Continental Journal of Microbiology* 3:27-30.

Oranusi, S. Madu, S. A. Braide, W. and Oguoma, O.I. (2011) Investigation on the safety and probiotic potentials of yoghurt sold in Owerri metropolis in Imo state, Nigeria. *Journal of Microbiology and Antimicrobials*: 3(6): 146-152

Oyeleke, S. B (2009). Microbial assessment of some commercially prepared yoghurt retailed in Minna, Niger state. *African Journal of Microbiological Research*: 3(5): 245-248

Sahan, N. Yasar, K. Hayaloglu A.A.(2008). Physical, chemical and flavor quality of non fat yoghurt as affected by a β-glucan hydrocolloidal composite during storage. *Food hydrocolloids*. 22:1291-1297.

Saint-Eve, A. Levy, C. Le-Moigne, M. Ducruet, V. Souchon, I (2008). Quality changes in yoghurt during storage in different packaging materials. *Food Chemistry* 110: 285-293.

Sera, M. Trujillo. J.A. Guamis. B. Ferragut, V. (2009). Flavor profiles and survival of starter cultures of yoghurt produced from high pressure homogenized milk. *International Dairy Journal*. 19:100-106

Shojaei, Z.A. and Yadollahi, A. (2008). "Physicochemical and microbiological quality of raw, pasteurized and UHT milks in Shahrekord". *Journal of Diary Science,* 59 (2):532-538

Sokolinska, D.C, Michalski, M.M, and Pikul, J (2004). Role of the proportion of yoghurt bacterial strains in milk souring and the formation of curd qualitative characteristics. *Bulletin of Veterinary Institute Pulawy*. 48:437-441.

Taura, D.W. Mukhtar, M.D, and Kawo, A.H (2005) Assessment of microbial safety of some brands of yoghurt sold around old campus of Bayero University, Kano., *Nigerian Journal of Microbiology,* 19(1-2): 521-528

Valera-Moreiras, J.M. Antoine, B. Ruyz-Roao, G. Valera, T. (1992) Effects of yoghurt and fermented, then pasteurized milk on lactose absorbtion in an institutionalized elderly group. *American Journal of Clinical Nutrition* 11:168-173.

Vinderolla, C.G. Bailo, J.A and Reinheimer, J.A (2000). Survival of probiotic microflora in Argentinian yoghurts during refrigerated storage. *Food Research International*.33:97-102

Younus, S.T. Masud. K and Aziz, T. (2002). Quality Evaluation of Market Yoghurt / Dahi, *Pakistan Journal of Nutrition,* 1(1): 226-230.

Plate 1:   Methyl-red test

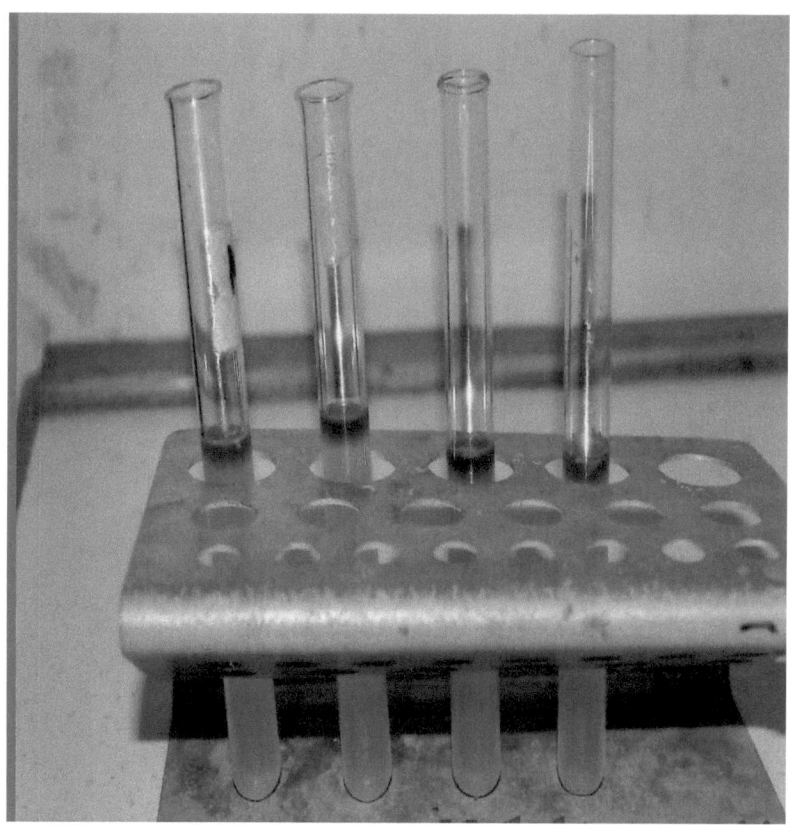

Methyl-red test - Positive

Plate 2: Indole - test

Indole test-Positive

Plate 3: Voges – proskauer test

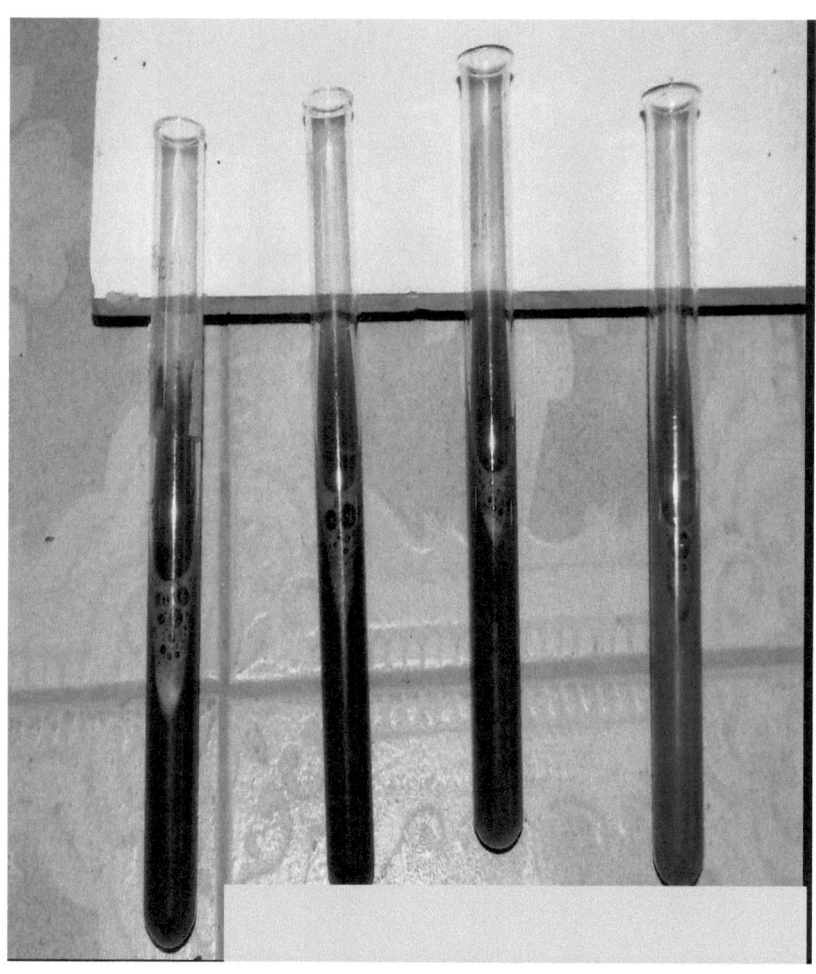

Voges-proskauer test - Negative

Plate 4:   Citrate utilization test

Citrate utilization test - Negative

Plate 5: **Photo micrograph of Mucor spp**

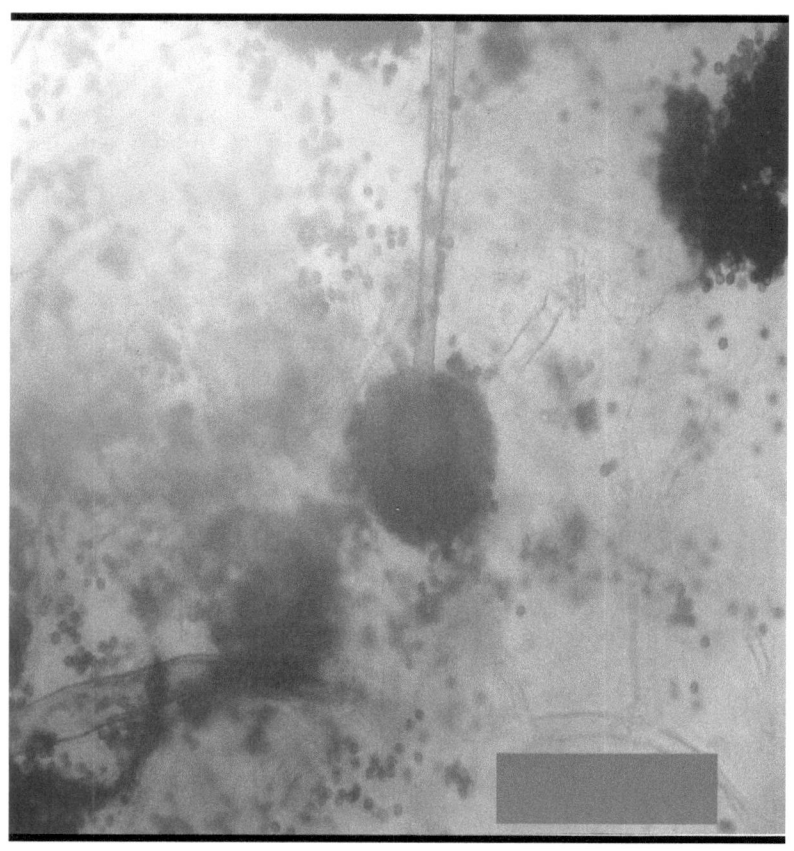

Photomicrograph of Mucor spp

Printed by Books on Demand GmbH, Norderstedt / Germany